W9-BOP-840

The Day of the Disaster

CHERNOBYL: NUCLEAR POWER PLANT EXPLOSION

Written By: Sue L. Hamilton

NOTE: The following is a fictional account based on factual data.

CHERNOBYL

Published by Abdo & Daughters, 6537 Cecilia Circle, Edina, Minnesota 55435.

Library bound edition distributed by Rockbottom Books, Pentagon Tower, P.O. Box 36036, Minneapolis, Minnesota 55435.

Library of Congress Number: 91-073040 ISBN: 1-56239-060-0

Cover Photo by: UPI Bettmann
Inside Photos by: Retna pgs. 3, 8, 11, 23, 31
UPI Bettmann pgs. 5, 16, 20, 25, 27, 30, 31

Edited by: John Hamilton

People marched to protest against nuclear fallout from the Chernobyl accident. The banner reading "Freedom and Peace" refers to the movement which organized the march.

FORWARD — MELTDOWN OF REACTOR 4

SATURDAY, APRIL 26, 1986
1:23 A.M.

The quiet of the night was shattered in the city of Pripyat, Soviet Union. This modern city, built to house and support 45,000 workers and families of the Chernobyl Nuclear Power Plant, would soon know tragedy.

Two thunderous explosions, seconds apart, marked the beginning of the worst nuclear plant accident in world history. Shrill screams of alarms shrieked out in warning. Within minutes, firefighters rushed to the scene on this early Saturday morning. Although well-trained to control and put out fires, this was a fire like no other they had ever known.

Here, at Chernobyl, which was named after an old-fashioned town only 11 miles away, a totally modern day disaster had begun — the meltdown of a nuclear power plant's core and the resulting release of life-threatening radiation into the air.

What caused this to happen? What does it really mean? What were the final effects? Follow the heroic actions of a Soviet firefighter as he describes to his wife the disaster whose invisible powers would reach out to affect the whole world.

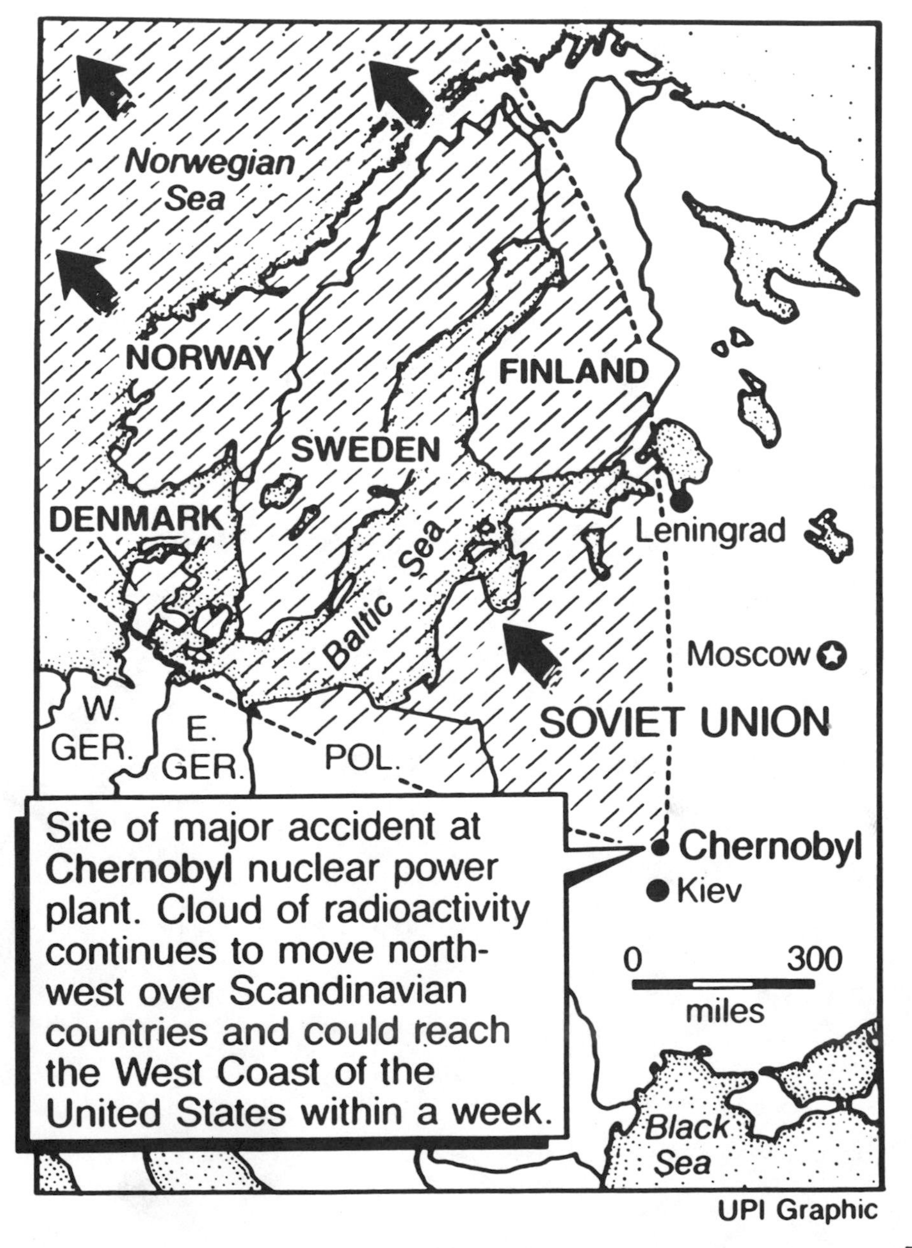

UPI Graphic

CHAPTER 1 — FIGHTING RADIOACTIVE FIRE

SATURDAY, APRIL 26, 1986
6:30 P.M.

I am in a hospital in Kiev using a tape recorder. Although I am ill and my fingers are so blistered and burnt that I am unable to write, I want to describe for you, Sasha, my wife, everything that I saw . . . all that has happened. I didn't feel good when I got here 12 hours ago — only half a day?! It seems like forever! Anyway, I am getting worse. Perhaps it is being in the hospital — I've never liked hospitals. But that is silly. You and I both know its the effects of radiation poisoning.

I am so glad that you were not here in Pripyat today. I had been missing you so much all week, but now I am thankful you were away visiting relatives in Moscow. The hospital people have been so busy, I do not know if you have been contacted yet. Still, it is just as well. I would not want you exposed to this, even here in Kiev.

It all began early this morning — around 1:30 a.m. Two explosions shook the whole area. I was sleeping . . . dreaming, and then for a second I thought we were having an earthquake. Things on the wall fell down, beds moved, pots and pans rattled, dishes broke, and just as I was beginning to really come out of my dream, the alarm rang. I was awake in an instant and joined with the others as we rushed out.

Quickly, we discovered the location of the fire — the power plant. It was a Code 3 — the most extreme emergency. We suited up immediately. All these years, no incident. Suddenly, today, two explosions — the worst thing imaginable had happened. As we took our places on the trucks, we knew it was Code 3. We knew the blasts meant things were bad, yet I don't believe anyone knew how serious the situation really was. None of us had ever been trained to know.

I'm sure your brother, the nuclear engineer, can explain to us what happened. But all we firefighters knew was what we saw. Our trucks arrived at the plant only a couple minutes after the alarm sounded. Although the night was pitch black, the nuclear plant was aglow — maybe it was the fire or the radiation, or both. But the plant gleamed a devilish red-yellow. If there had been time, perhaps I would have been afraid, but there was so much to be done, we just got right to work.

Firefighters work through the night to control the deadly inferno.

Lieutenant Pravik — you remember Vladimir don't you? — saw right off that we'd need more help. Our three fire engines and crew weren't going to take care of this mess. He called for additional support immediately. Fire brigades were brought in from around Pripyat, as well as Chernobyl and Kiev.

The first order of business was the collapsed, flaming roof on top of Reactor 4 — which I now know was where the actual meltdown occurred. One team went into the building, and another went up on the roof to tackle those fires. I would already be dead had I been among them. Pravik went in there. I understand he has since died.

Major Leonid Telyatnikov took over command. He led a group of us up to the roof of Reactor No. 3. There were five fires there — we had to get them out to keep the other buildings from going up.

The fire was intense. So hot, in fact, the roof was melting. Each step coated our boots with a hot, sticky tar, making them heavier and heavier. It took all my strength just to get over to the fire.

The fire itself was not like the heat of an ordinary house fire. It was more of a searing feeling. I know about the existence of radiation, but what is strange is that it's invisible. It is one thing to fear something you can see — a fire or a tornado — but this was something different. Unseen, but there. And yes, we did feel it.

Everyone worked professionally, consistently. We had no special protective gear, no respirators, no training for handling a fire at a nuclear plant, and no equipment for measuring radiation levels. Why is it so obvious to us today that these things were so critically needed, now that the tragedy has passed? Why must people die for things to be put right?

Anyway, we fought the fires as best we could — using water and hand-held extinguishers. It was hard to see. The water we used to put out the fires turned to steam — blisteringly hot radioactive steam. We continued to inhale the radioactive particles, continued to be burnt, but the fires had to be put out. It was our job. Our duty.

I was beyond exhaustion as I climbed down from the roof. I had a strange metallic taste in my mouth and a severe headache. Blisters covered my hands, face and back. I was sick to my stomach, but the fires were out. Not much left for us to do.

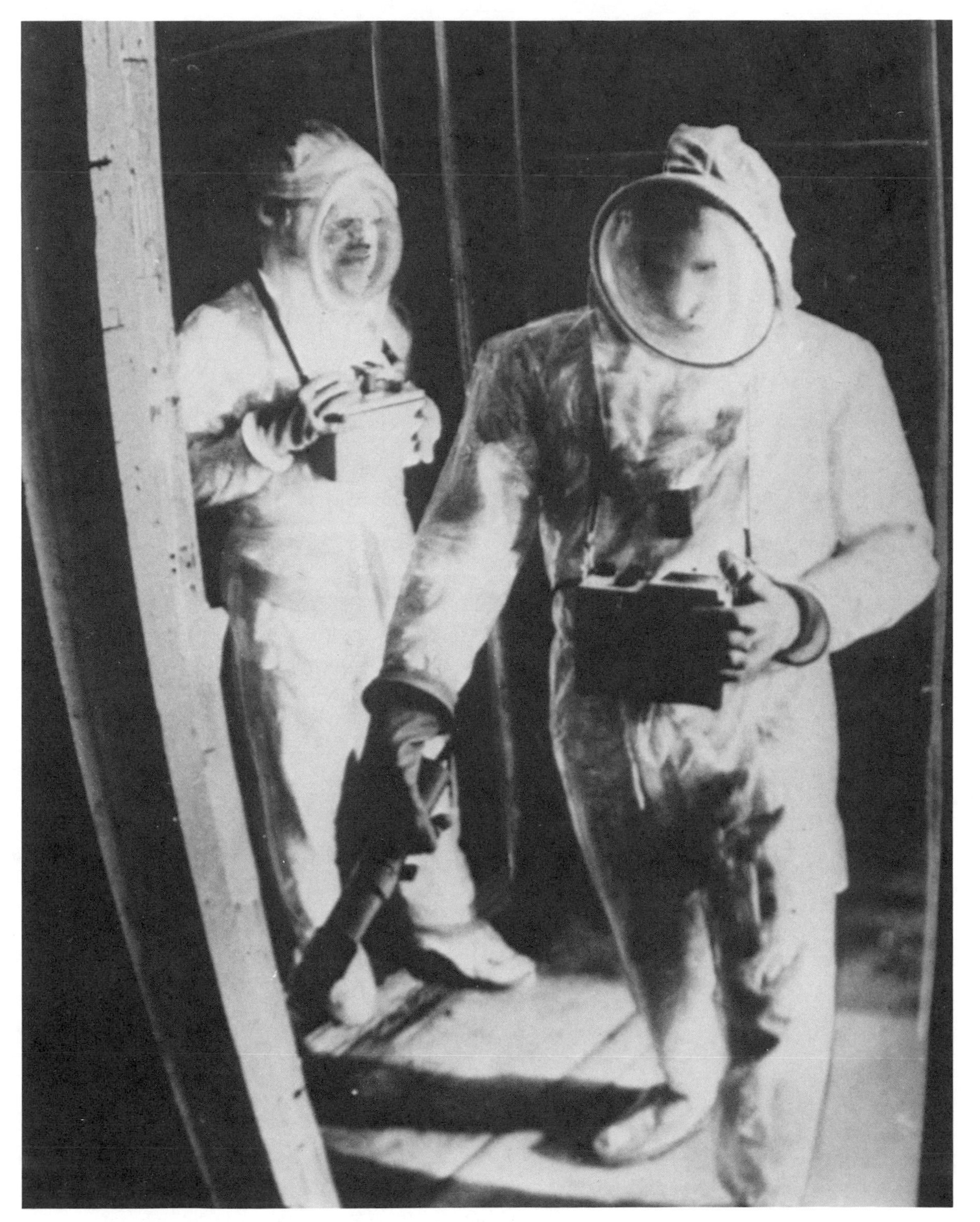

Radiation experts, wearing protective gear, test the area for levels of the invisible killer.

I was assisted to a stretcher, where I promptly collapsed. Immediately, I was carried toward a waiting ambulance parked just out of the way. Away from the main firefighting effort, I looked up in the sky. The stars shown. A gentle breeze blew through nearby trees. Should it be like this, so beautiful, when all around us disaster loomed? Then I looked closer at the trees. Pine trees that had stood solid and green only hours before had begun to yellow and die. That showed the clear effects of radiation on living things. Suddenly, I was scared and wanted to leave. I did my job to save others, Sasha, but I don't want to die like the pines!

As they put me in the ambulance, I heard the sound of helicopters and asked what was happening. They were taking up loads of sand, boron and lead shot to be dropped into Reactor 4, where the meltdown was still raging. As I struggled to control my fear, I suddenly realized that the worst was not over. I think my few hours at the plant may have helped others, but it was only the beginning.

CHAPTER 2 — HOSPITALIZED

SUNDAY, APRIL 27, 1986
1:30 P.M.

I continue to have blood transfusions. I do not believe any of my own blood is still in me. Everything in me or on me has to be carefully disposed of. All my clothing was removed, bagged and stored for disposal. (I presume it'll be buried or burned.) The same is true of my blood. Even when I go to the bathroom — that must be properly disposed of, because it is radioactive.

Everything around me is sterile — clean as clean can be. To reduce the spread of radiation, the nurses and doctors wipe their feet on damp towels before entering or leaving the room. I have been carefully washed. My skin is quite blistered. My hair is falling out and my head still aches. Still, I do not believe I am as bad as some others. There is a group scheduled to fly to Moscow tonight for special emergency care — bone marrow transplants.

When my doctor came in, I asked him to give me a little information to go with my iodide pill. Specifically, what is radiation? What does this pill do. Am I OK? Do I need a bone marrow transplant?

He was very straightforward. Radiation, in simplest terms, is made of particles or waves given off by an unstable atom. Do you remember your brother's explanation of how Chernobyl worked? In a nuclear power plant, uranium atoms are broken apart to achieve great heat. The heat is cooled by water, which creates steam, which spins a turbine, which in turn generates the electricity we use. Well, a great deal of unstable atoms are created as the uranium atoms are broken apart. If everything goes as it's supposed to — as Chernobyl did since 1977 — no radiation leaks out of the contained area. The plant generates clean electricity. Still, something happened and the explosion occurred, opening the plant to the outside world and shooting radiation into the air.

The doctor pointed out that we are surrounded by natural radiation all the time. From the sun, radioactive rocks deep inside the earth, even from plants, animals and other humans. All these things give off tiny amounts of radiation that our bodies are used to handling. But I was bombarded by a much greater dose. I have been exposed to anywhere from 200 rems - 500 rems. A rem is a unit of measure of absorbed radiation.

Exposure to thousands of rems will kill you within hours or days (probably what Lieutenant Pravik experienced.) From 700 to 1,200 rems, death occurs within days or weeks from internal damage. From 300 to 700 rems, a bone marrow transplant is needed because your own bone marrow is so damaged it can no longer make white blood cells, which protect the body from infection. Exposure of 200 to 400 rems kill half the people within 30 days. 100 to 200 rems injure the immune system and cause a long-term risk of cancer. Less than 100 rems makes you sick to your stomach, although the sooner you get away from exposure, the less chance you'll have of getting cancer.

The iodide pill is taken to fill my thyroid gland — the gland that produces hormones and controls growth and development. Apparently, the thyroid takes in as much iodine as it can. One of the main products from the nuclear explosion was something called iodine 131. If one's thyroid is full, it will not absorb the radioactive iodine 131. Thus, the pill.

Overall, I do not like my chances. The doctor says that my exposure has not been as bad as others but it is still bad. I am coughing, sick. Still, how could something you can't see do this? I refuse to believe I will die. I'm only 26 years old. I'm young! I'll fight it!

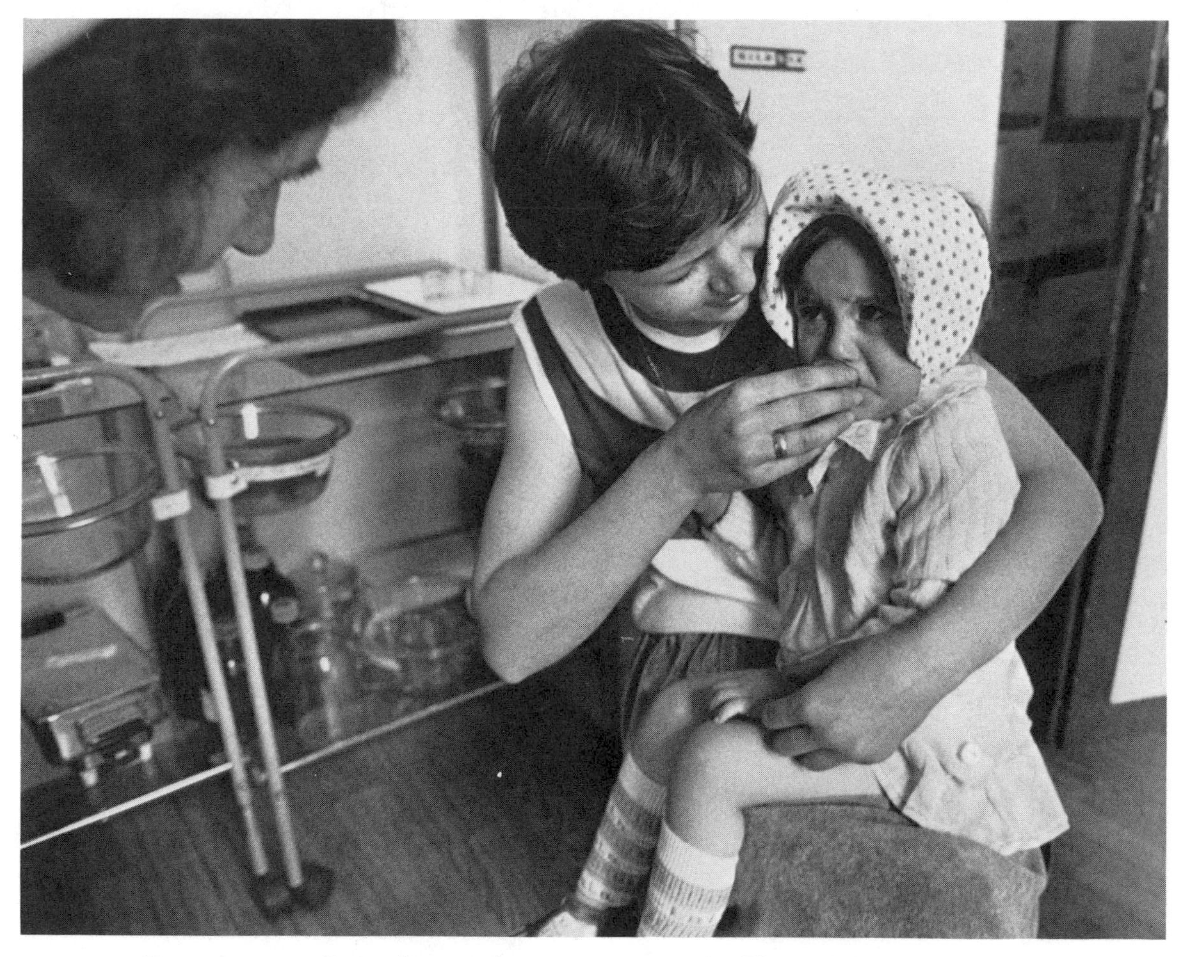

Exposure to radiation has made everyone near the Chernobyl plant very sick.

CHAPTER 3 — WHAT HAPPENED?

MONDAY, APRIL 28, 1986
5:30 P.M.

Alek, your brother, visited me today. I did not recognize him at first. I thought he was just another doctor. He had on a protective gown, gloves, and coverings on his shoes. Everything is white. How I wish to see another color! Not red, though . . . I still think of the glow of the power plant when I think of red.

Alek said he had a hard time finding me. I had been transferred almost right away here to Kiev, and what with all the cases in the hospital in Chernobyl, it took awhile to locate me. He told me that you wanted to come here, but he convinced you to stay in Moscow. I am glad. I do not want you exposed to this — although I miss you.

I asked him what had happened. He was hesitant at first to tell. He said he had heard only rumors. I told him I wanted to know. I was sick from this, and thought it was my right. Again he hesitated, then I saw it in his eyes — a combination of pity and sadness for me. Step-by-step, he pieced together the steps to disaster:

Step 1 — Chernobyl technicians decided to conduct a test to see how long the steam-driven turbines at the plant would continue to generate electricity if there was a power cutoff.

Step 2 — Friday afternoon, at 2:00 p.m., the technicians purposely shut off the plant's emergency cooling system. A very dangerous move — this meant there was no backup should the plant begin to overheat, but they thought they were in control and needed to do this to conduct the test.

Step 3 — A dispatcher did not want to shut the plant down, but insisted on continuing to generate electricity for the area. Against safety regulations, the reactor continued to run at 50% power for the next nine hours with no emergency backup cooling system.

Step 4 — As the hours passed, a radioactive gas — xenon — began building up in the reactor. That should have put a stop to the experiment, as the gas causes the reactor to become unstable and lowers its power output. But they continued.

Step 5 — Down to 1% of its capacity, the technicians needed to build up the power. To do this, they removed control rods and disconnected the automatic-rod-control system, which would have helped them shut down the reactor in case of an emergency.

Step 6 — Now they saw that the water level was too low, and turned on additional coolant pumps. This brought too much cold water into the system and caused the hot steam to form water droplets, which further destabilized the reactor.

Step 7 — With water and pressure levels changing, the technicians were afraid that the system would shut itself down — as it was supposed to. To complete their experiment, they purposely blocked the emergency water- and pressure-level warning signals.

Step 8 — At 1:23 a.m., the experiment began as the technicians shut off valves to prevent steam from reaching the turbine unit they wanted to test. Alek said that last step sealed the reactor's doom. Normally, this would have caused the reactor to — as he put it — "scram," to shut down immediately. But they didn't want to restart the reactor if they had to do the test a second time, so they bypassed the warning shut-down signal, too.

At the center of the reactor, workers monitor the pressure that builds up.

Heat and steam built up immediately inside the reactor. Alek guessed that it took only seconds for them to realize the true danger they had created. They tried to put the control rods back in. These rods control the reactor's energy release rate by acting as sponges and absorbing the neutrons that would cause other atoms to split. Thus more rods inserted into place, the less fission is created. Had they been properly in place, they would have stopped the inevitable reaction. But it was too late.

The first explosion was probably from the build-up of steam and pressure, rupturing the tubes in the reactor. The second, a hydrogen-induced blast, blew the top off the building. This exposed the graphite core. As a 4,000° Fahrenheit fire began to burn, radioactive particles blasted up into the sky.

This is the fire I went to fight with standard firefighter's gear.

CHAPTER 4 — THE AFTERMATH

TUESDAY, APRIL 29, 1986
5:45 P.M.

I am getting weaker each day. Radiation is like a poison; it spreads and worsens each day. I may be moved to Moscow, near you, Sasha, afterall.

Today I asked Alek what was being done for the people of Pripyat and Chernobyl. He said the entire area for 18 miles surrounding the power plant had been evacuated. I was thankful . . . then disgusted to learn that the evacuation did not take place until Sunday! For an entire day, mothers, fathers, children continued to live in this area, bombarded with radiation! What will happen to them, I asked my doctor? He shrugged. He does not know for sure, but many will get cancer. Many will die.

One good thing was that the meltdown happened at night. Had it occurred during the day, over 2,000 people would have been working on the construction sites of Reactors 5 and 6. As it was, less than 500 construction and staff people were there. People were at least partially protected from radiation by being inside their houses. Little comfort for these people, but still something.

The area around the nuclear plant has been evacuated and workers monitor the levels of radiation.

I must admit it is quite a feat to have transported 45,000 people to new locations. Alek says that Pripyat is now a ghost town. On Sunday morning, residents were given four hours to put their things together. Now, not a soul remains. Only the pets, who weren't allowed on the buses, roam the streets. They will die soon, I am sure. Everyone, via hundreds of buses, has been taken to live in other areas. They will be given new jobs. If I live, I will be a firefighter somewhere else. Pripyat will be uninhabitable for years. It is unlikely that these people will ever return to their homes. This is probably true for other areas, as well. How tragic to leave all you know to go someplace strange. But what else can be done?

Outsiders have not been informed of the disaster. It is an embarrassment for the Soviet people. However, if radiation can travel on the winds, as I've been told, it will spread across Europe, endangering others. I think people should be told . . . be warned. No one should have to go through what I am going through — the weakness, sick to my stomach, headaches and burns.

I do not know how much longer I will go on, Sasha. But I do know that this disaster cannot be ignored. What a high price we have paid for electricity and power. Something must be done!

******

NOTE: This fireman lived, although he continues to struggle with the effects of radiation poisoning — weakness, trouble breathing, and high risk of cancer. If he has children, they are at risk of being deformed, and will also have a high risk of cancer.

The firemen at the Chernobyl plant were honored as heroes of "The Battle of Chernobyl." Many died — even those with special bone marrow transplants did not live. However, their brave efforts to put out the fires and to contain the area saved thousands of people from additional exposure to the killer radiation.

Weeds are quietly taking over the pavement in the town of Pripyat, near Chernobyl nuclear power station. Officials say the deserted town will remain abandoned as nuclear cleanup would prove too costly.

CHAPTER 5 — NUCLEAR POWER'S VAST RISKS

Chernobyl was the worst nuclear power plant disaster in the history of the world:

- 31 people were directly killed by the accident — although estimates have gone as high as 300.
- Over 135,000 people had to be evacuated from 179 villages and towns in an 18-mile radius around the plant.
- Radioactive particles carried by the upper atmosphere drifted over northern Europe, causing radiated rain to drop on unwarned communities in Norway and Sweden.
- Throughout northern Europe, 400 million people were exposed to the radiation.
- $358 billion dollars were spent for cleanup and the cost of lost farmland, crops and animals.
- Serious health conditions developed, including dramatic increases in thyroid disease, leukemia, blood disease, anemia and cancer.
- Large areas surrounding the reactor site in the Ukraine and in nearby Belorussia have documented an increase of deformed animals and plants.

People continue to take precautions against the fallout. In this picture a youngster swallows an anti-radiation iodine solution in a Warsaw clinic.

Today, Chernobyl is still on the minds of millions of people. In Kiev, 80 miles south of Chernobyl, milk, fruits and vegetables are all constantly checked for radiation levels. People are also checked. They must wash every day, leave clothes and shoes outside. Windows are closed tightly. Cars are checked and washed. Roads are kept wet in order to keep down radiated dust. People must keep away from leafy trees such as poplar or chestnut when it rains, as the rain could pick up radiated dust and drop it on their heads.

Many areas closer to the nuclear plant are still closed off to the outside world. Everything — the soil, plants, buildings — is contaminated with radiation. It will be years before it is safe once again.

The Soviet Union, with 50 nuclear plants, is not the only country to experience this type of disaster. The United States, leading the world with over 100 operating reactors, saw disaster in 1979 at Three Mile Island near Harrisburg, Pennsylvania. Here again, the water coolant did not reach the reactor, and a partial meltdown occurred. Design differences kept this disaster contained and did not result in an open meltdown, as was faced in Chernobyl. In general, United States nuclear plants have been regarded as much safer then those operating in the Soviet Union.

Many other large countries around the world operate anywhere from 2 to 39 reactors. However, no matter how safe nuclear plants profess to be, there are great risks. The threat of equipment failure or human error (as in Chernobyl) is always present. Scientists estimate a meltdown will occur somewhere in the world every 20-30 years. It is likely that each generation will face a nuclear disaster.

On the positive side, nuclear power, when it works the way it should, is clean, quiet, and does not require the burning of fossil-fuels — such as coal or oil — which are great causes of air pollution. A national organization called the World Association of Nuclear Operators has been created to exchange information on the running of power plants. Additional safety measures have been presented and agreements have been made with all nations to see that no secrets are kept regarding nuclear accidents.

There will always be risks with nuclear equipment. However, we have come to depend heavily on its power. All we can do is enforce the strictest safety and design measures and hope that Chernobyl is our last nuclear day of disaster.

Some 20,000 people demonstrated in Hamburg to remember the anniversary of the Chernobyl nuclear tragedy.

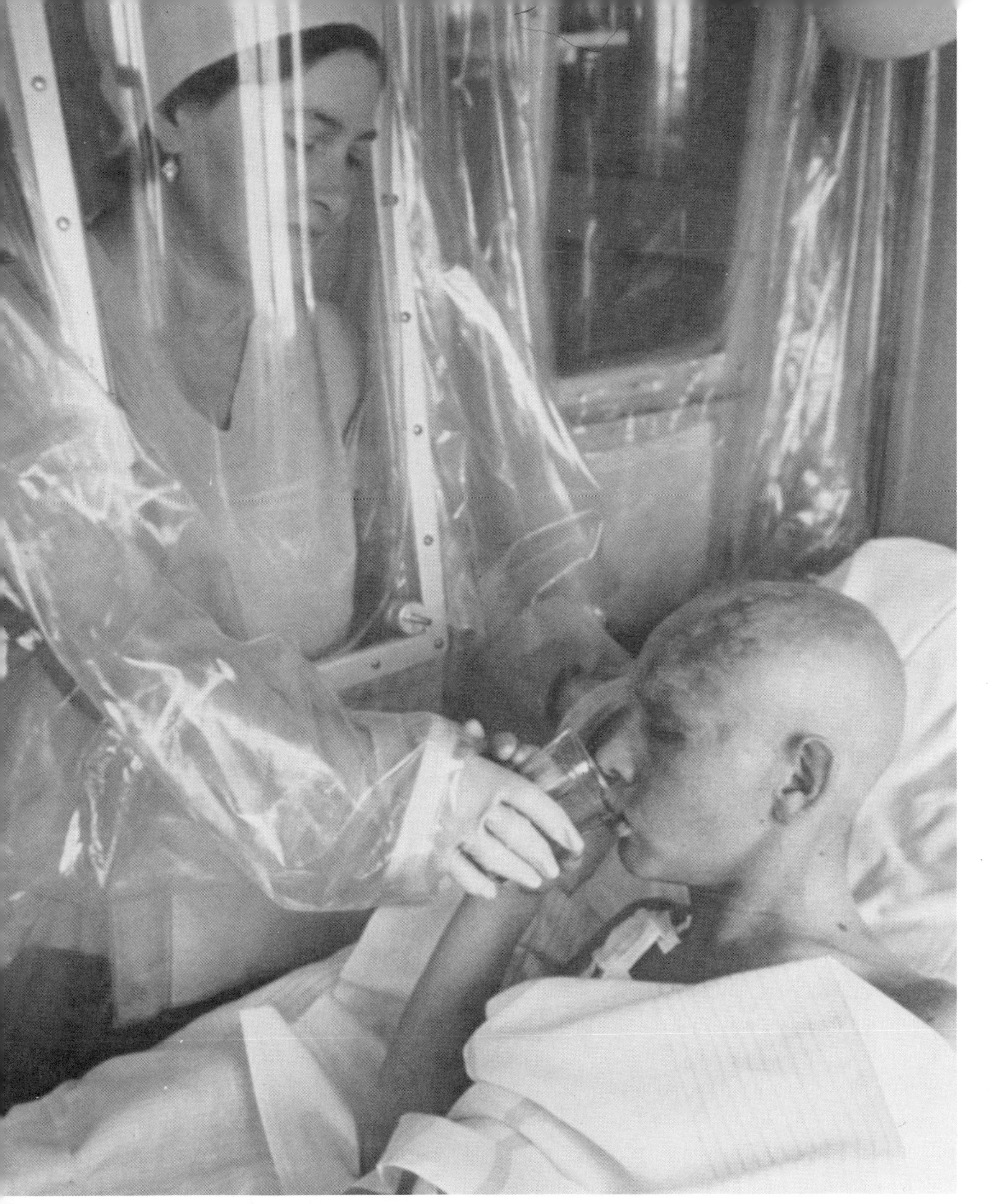

People near Chernobyl continue to feel the effects of the nuclear disaster.

Lt. Colonel Telyatnikov, head of the fire brigade which brought the Chernobyl nuclear power plant fire under control, points out the damage done to the fourth reactor during the April 26, 1986 nuclear disaster.